W9-CMQ-372

# Making a Clock-Accurate Sundial

## Customized to Your Location

### (for the Northern Hemisphere)

## by Sam Muller

JUN    '00

HAZARD BRANCH
Onondaga County
Public Library
1620 W. GENESEE STREET
SYRACUSE, NY 13204

Naturegraph Publishers

**Library of Congress Cataloging-in-Publication Data**
Muller, Sam,1939-
    Making a clock-accurate sundial : customized to your location
for the northern hemisphere / by Sam Muller.
        p.   cm.
    Includes index.
    Summary: Presents step-by-step instructions for making a
sundial which will illustrate concepts regarding the interrelation
of the sun, the earth's rotation, and orbit, and time.
    ISBN 0-87961-246-0 (alk. paper)
    1. Sundials—Design and construction—Juvenile literature.
    [1. Sundials—Design and construction.] I. Title.
    QB215.M78  1997
    681.1'112—dc21                                             97-8226
                                                                                    CIP
                                                                                    AC

Copyright © 1997 by Samuel A. Muller.

All rights reserved. Printed in the United States of America. No part
of this book may be used or reproduced without prior written
permission of the publisher.

All illustrations by Samuel Muller

Naturegraph Publishers has been publishing books
on natural history, Native Americans,
and outdoor subjects since 1946.
Please write for our free catalog.

Naturegraph Publishers, Inc.
3543 Indian Creek Road
Happy Camp, CA 96039
(916)493-5353

Books for a better world

# Table of Contents

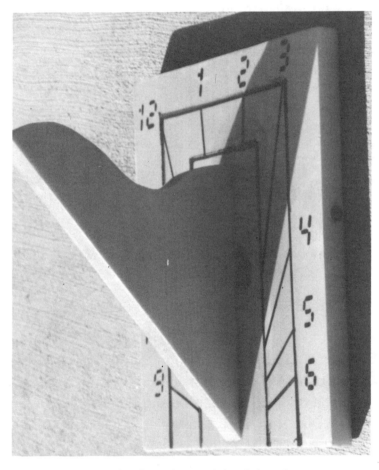

A completed wooden sundial with hour lines

# Introduction

The modern sundial will work in the Northern Hemisphere north of the equator. I call it modern because the method for making it is new, different, and easy. It is easy to make because the method bypasses all complicated mathematics. Also, very little woodworking skill is required. It is a sundial custom-designed for your spot on the earth by you; the only way you are likely to have a sundial like this is to make it yourself.

When done, you'll have a sundial that is always close to clock time. For exact time you will need to add or subtract a few minutes for the particular day using the exact correction chart on page 36 (see Chapter 4).

The predictable slight difference in sundial time is a result of the earth's tilt and its eccentric orbit around the sun. Whenever you look at the sundial's shadow, besides reading time, you are indirectly observing the dynamics of the earth's motion.

You can also use this sundial to observe lunar motion during the cycle of the full moon by watching the position of the moon's shadow on the sundial. By taking a sundial like this along on a trip, you can observe yet other interesting phenomena explained in Chapter 5.

I think that when you have completed reading this book and constructed your own sundial, you'll agree that the modern sundial is fascinating, educational, and fun!

# Chapter 1

# A Cardboard Sundial

Before building a sundial out of wood, it's necessary to build a small, working model out of thin cardboard. With just a little time and effort you'll learn some important things about sundials quickly and easily that will help you later (in Chapter 6) to understand some basic concepts about the earth's rotation, the sun, and time zones.

This cardboard sundial is a lot like the one you'll make of wood. It has just two parts. They are the gnomon (pronounced *NO-mun*) and the plate. The gnomon is a triangular-shaped part that casts the sun's shadow. The plate is a rectangular-shaped part on which the shadow falls. Time is read from the position of the shadow on the sundial face, the name for the top of the plate.

You likely have all the materials right at hand you need to make this sundial. You can make it out of posterboard or the cardboard of cereal boxes (like cornflakes, etc.). Here is a list of the materials you will need.

**Materials for a Cardboard Sundial**
1. sheet of cardboard
2. protractor
3. pencil

4. scissors
5. masking tape
6. white glue
7. measuring stick
8. adjustable magnetic compass
9. geographical atlas
10. topographical map of your area

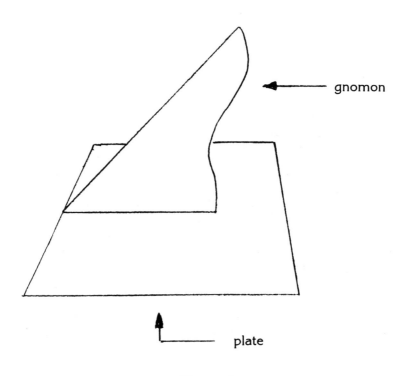

Diagram 1

## Making the Gnomon

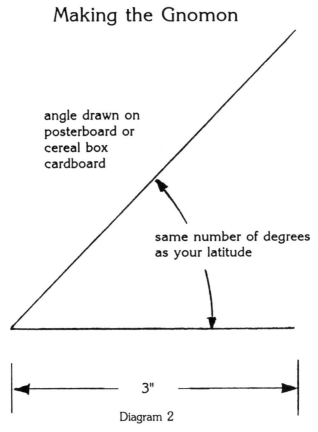

angle drawn on
posterboard or
cereal box
cardboard

same number of degrees
as your latitude

3"

Diagram 2

First you'll make the gnomon. On the cardboard draw an angle with a protractor that has the same number of degrees as your latitude. Your latitude is the amount of angular distance you live north from the earth's equator. This distance is measured in degrees from zero to 90, zero degrees being the equator and 90 degrees the North Pole. For example, someone living in Columbus, Ohio, would draw a 40 degree angle, because Columbus is located about 40 degrees north from the equator. You can determine your latitude by looking at the latitude lines on a map of your location in an atlas. Make the length of the base of the angle 3 inches. The other side of the angle should be just a little longer. Its tip should be just above the end of the base.

Connect the ends of the two sides of the angle with a curved line—something like a stretched out, backwards S. This will make reading the time easier. Read time only from the shadow that is a straight line; never read time from the curved shadow.

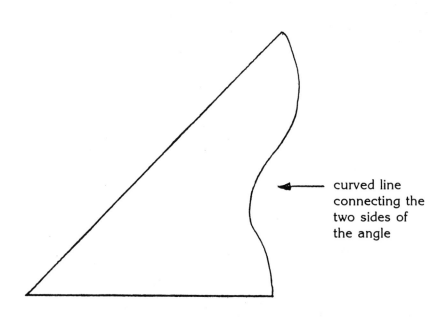

curved line connecting the two sides of the angle

Diagram 3

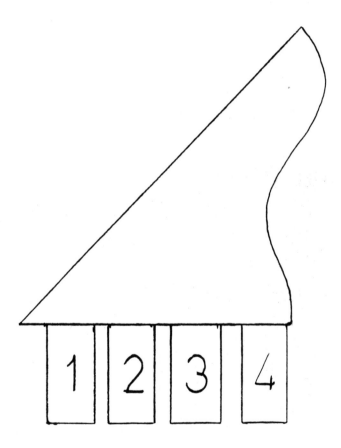

Diagram 4

On the bottom of the 3-inch base of the gnomon draw four tabs, ½ inch by 1 inch. Number
them 1, 2, 3, and 4.

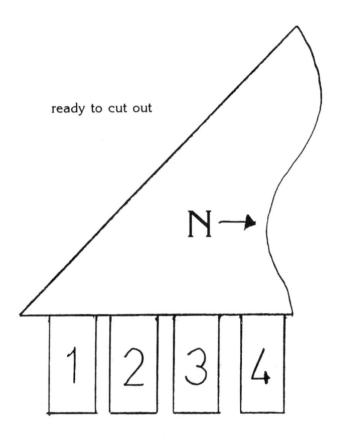

ready to cut out

N→

Diagram 5

Put an N and an arrow on the gnomon with the arrow pointing to the side of the gnomon formed by the curved line. Then cut out the gnomon with a pair of scissors with its numbered tabs attached to it.

# The Sundial Plate

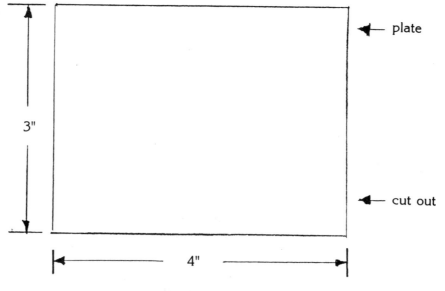

Diagram 6

Now draw a rectangle on the cardboard. Make it 3 inches by 4 inches. It is very important that the sundial plate be perfectly flat. If it is warped, it will lead to slightly incorrect readings. When you have drawn the lines for the sundial plate, cut it out.

# Joining the Sundial Plate and Gnomon

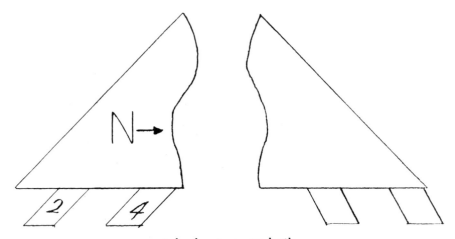

tabs bent up on both
sides of the gnomon

Diagram 7

Attach the gnomon to the sundial plate in the following manner. Bend the even-numbered tabs up so they are on the right side of the gnomon. The bent tabs will form a 90 degree angle with the side of the gnomon. Bend the odd-numbered tabs up so they are on the left side of the gnomon. They also will form a 90 degree angle with the side of the gnomon.

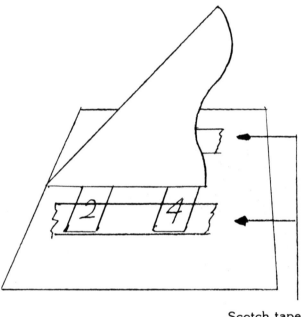

Scotch tape

Diagram 8

Attach the gnomon to the middle of the sundial plate with masking tape (or Scotch tape)
over the tabs. Don't let the tape quite touch the
bottom of the gnomon because you'll be putting
a little glue along the base to make it stronger.

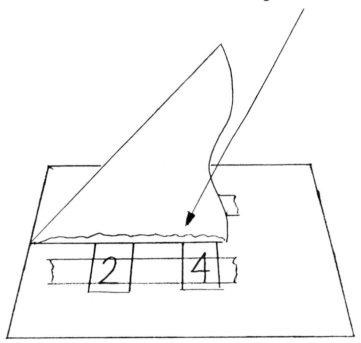

Apply white glue along
the bottom of the gnomon

Diagram 9

Lay down a line of white glue along the base of the gnomon where it meets the face. Put glue along both sides of the bottom of the gnomon. Let it dry.

# Using the Sundial

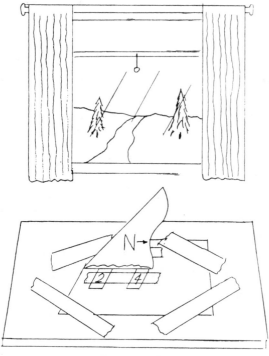

Diagram 10

When the glue is dry, you'll use this sundial indoors to get an idea of how it works. Put it on a windowsill or counter where it will get sunlight for part of the day. The spot selected must be level and stationary. Position the sundial on the sunny spot so the gnomon points to true north. Use an adjustable magnetic compass to find true north. Tape the sundial securely in that position with masking tape. The sundial will stay there for a few days. It would be preferable to have the sundial taped to a Formica surface, marble windowsill, or similar surface that won't be damaged by removing the tape some time later.

Although finding true north is not so important at this stage, and simply using the magnetic north pointed to by your compass needle is sufficient, it is essential to find true north when making the permanent sundial described in the next chapter. The difference between true north and magnetic north is called magnetic declination, which is caused by distortions in the earth's magnetic fields. To adjust your compass for magnetic declination you need to know how many degrees east or west of true north your compass is pointing. In the United States and Canada, the north and magnetic poles happen to be in line along a narrow strip that passes through the Great Lakes. At all places east of this line, the compass needle will point west of true north, and at all places west of this line the compass needle will point east of true north.

You can find the declination setting for your location by consulting a topographical map of your area, where it will be written in degrees east or west of the true north line, or by calling a local surveying office. How to adjust your compass for magnetic declination should be explained in the instructions that came with your compass. Basically, it involves turning the compass dial the number of degrees required in the opposite direction of your declination. For example, say you are in eastern Canada at a location where the declination is 15 degrees west. This means the compass needle is pointing 15 degrees too far west. You place the compass down (not near any metal objects or close to any electrical appliances) and rotate the compass dial to your right 15 degrees from zero until the bearing marker on your compass is at 345 degrees. Align the compass needle with the bearing mark, and your bottom arrow inside the capsule will now be pointing to true north.

You can also find true north if you know how to find the North Star. It is always within one degree east or west of true north, since the North Star revolves around the North Pole. To find the

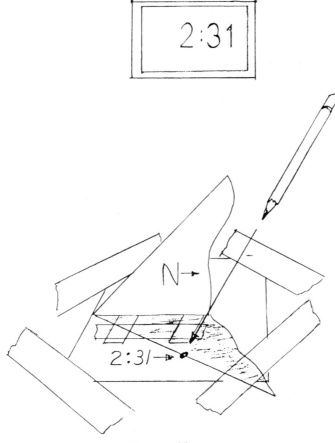

Diagram 11

North Star, look for the Big Dipper. The two stars forming the front of the Dipper's bowl, called "pointers," always point toward a bright star nearby, which is the North Star.

When the sun is shining on the sundial, mark a few of the positions of the shadow. All you do is put a dot anywhere along the edge of the straight shadow cast by the gnomon. Next to the dot record the clock time and draw an arrow from the clock time to the dot. At random, mark several times in that way.

If the sun is shining the next day, this is what you will see. You'll find that the sundial is keeping just about perfect time; it'll be doing a better job than a lot of clocks! As the shadow returns to each dot, the time you recorded the day before for that dot and the clock will agree.

If you leave the sundial in that position for some days, eventually you will see that it is gaining or losing a little time compared to the clock. In the fall or winter, it may not take too many days for you to see this happen. In the spring or summer, it may take quite a while before you notice it.

Basically, you've built an accurate timepiece and a simple scientific instrument. Obviously, if the sundial face were bigger, you could write in more times in an orderly way and have a complete sundial. That's what you will be starting to do in the next chapter. **Save this cardboard sundial because you will need it later.**

# Chapter 2

# Making a Clock-Accurate Wooden Sundial

Now you will begin to make a simple, wooden sundial. It will be similar to the cardboard one you've already made. You'll be designing it to work well on your particular spot on the earth where you will use it. You'll need a good pine board, 1 inch by 10 inches by 6 feet. This is more wood than you really need, but 6 feet is often about the shortest length you can buy. In case you make a mistake, you'll have plenty of extra wood. Also, later you may decide to build more sundials. Here is a complete list of materials needed to make a permanent modern sundial.

**Materials for a Wooden Sundial**
 1. pine board 6 ft. x 10 in. x 1 in.
 2. protractor
 3. pencil
 4. regular saw & coping saw
 5. sandpaper
 6. wood glue
 7. ruler
 8. permanent marker
 9. two #10 or #12, 2 inch flathead woodscrews (optional)
10. wood drill and bits (optional)
11. polyurethane wood preservative

12. adjustable magnetic compass
13. geographical atlas
14. bubble level

## Making the Gnomon

On the board draw an angle that has the same number of degrees as your latitude. Make it look like diagram 12. Note that the base of the angle is 8 inches long. It runs in the same direction as the grain of the wood. Also, the side of the board forms the base of the angle. This assures the bottom of the gnomon will be perfectly flat and fit nicely on the sundial face. Connect the ends of the two lines forming the angle with a wavy line. It's a line that looks like it was straight but got pushed in or bent. Its purpose is

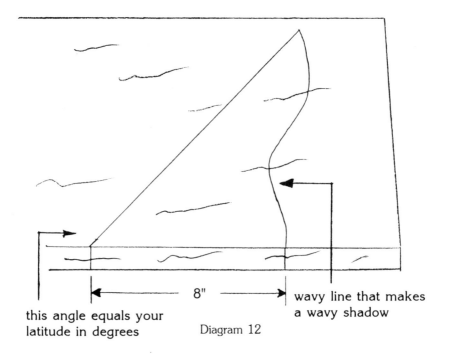

8"

this angle equals your latitude in degrees

wavy line that makes a wavy shadow

Diagram 12

to make sure you don't read that part of the shadow to get the time.

When your gnomon resembles the one in diagram 12, cut it out carefully. You can easily do it by hand with a regular saw for the straight edge and a coping saw for the curved edge. Sand it to remove splinters. The cut-out gnomon should resemble diagram 13.

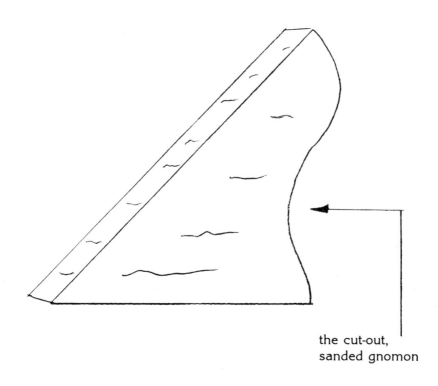

the cut-out, sanded gnomon

Diagram 13

## Making the Sundial Plate

Now draw a rectangle on the board for the sundial plate. The actual width of the board will be one dimension of the sundial plate. Make the other, longer dimension (the one that runs in the same direction as the grain of the wood) $11\frac{1}{2}$ inches.

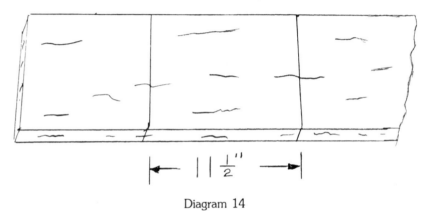

$$\longleftarrow 11\frac{1}{2}'' \longrightarrow$$

Diagram 14

When the board resembles diagram 14, cut it out with a saw, and sand off the splinters. Test the plate to make sure it is not warped by placing it on a flat surface. Press down on the corners and release your fingers to see if it wobbles. If it wobbles, turn it over and test the other side. This usually solves the problem. The bottom of the sundial must not wobble when placed on a flat surface; otherwise the time reading could be off by as much as five minute when permanently mounted in a level position.

## Joining the Gnomon and the Plate

Center the gnomon on the rectangular plate in the same position as you did for the cardboard sundial. Be sure the back edge of the

gnomon and the plate line up evenly before attaching the gnomon to the plate with wood glue.
Put the glue on the bottom of the gnomon. The
grain of the wood should run in the same direction on both the sundial plate and the gnomon
when they are joined. Apply a little hand pres-
sure to the gnomon for a few moments. Wipe off
the excess glue quickly with a damp cloth. Let it
dry.

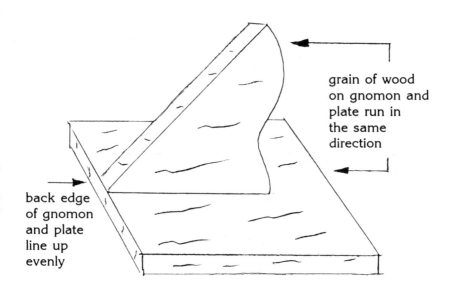

grain of wood
on gnomon and
plate run in
the same
direction

back edge
of gnomon
and plate
line up
evenly

Diagram 15

The glued-together sundial will resemble diagram 15. If you wish to make the sundial extra strong, put two #10 or #12, two-inch flathead woodscrews through the bottom of the sundial plate and into the bottom of the gnomon. Predrill the holes, countersink them, and put some glue on each screw.

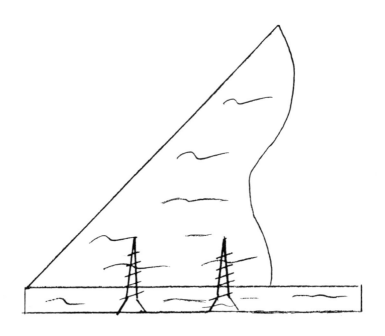

Diagram 16

## Coating the Sundial

Sand and clean the sundial. Then give it several coats of clear, interior/exterior polyurethane coating. Use either a brush or a spray can. Follow the directions that accompany the product.

When dry, put an N on the front of the sundial. Make it the same color as you're planning to use for the hour lines and the hour numerals. Permanent markers are all right. Paint sticks are better because they don't fade nearly as fast when exposed to the weather.

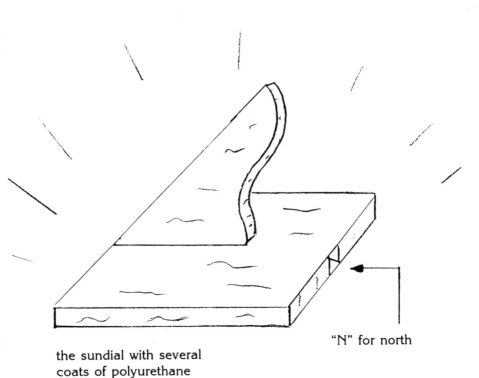

the sundial with several
coats of polyurethane

"N" for north

Diagram 17

## Preparing an Area for the Hour Lines

Now you are ready to draw a block U-shaped area on the sundial face, where lines indicating hours, half hours, and quarter hours will be drawn. First, using a ruler and a pencil, draw a line around the gnomon that is 1¼ inches from the sides of the gnomon and almost touches the gnomon in front (i. e., about ⅛ inch away).

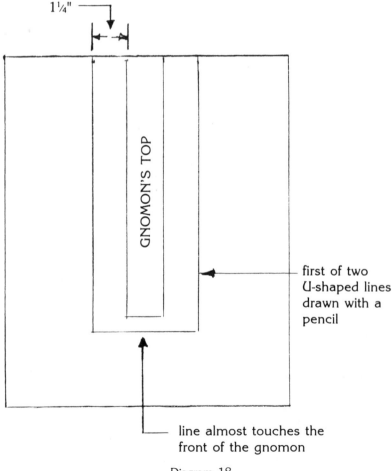

1¼"

GNOMON'S TOP

first of two
U-shaped lines
drawn with a
pencil

line almost touches the
front of the gnomon

Diagram 18

Then draw another, slightly larger capital U, 1½ inches away from the first one. These two lines enclose a narrow area where the hour lines will be drawn. The area outside this section will be for putting in the hour numerals. After making the lines with a pencil, trace over them with a marker.

Top View of Sundial

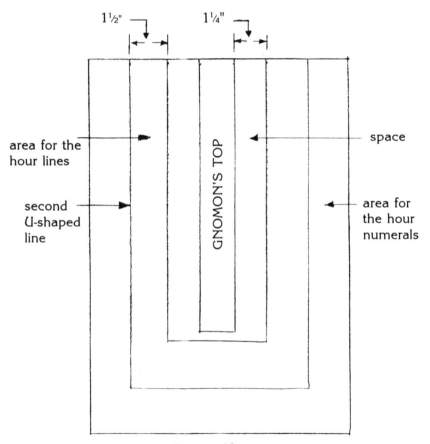

Diagram 19

Before putting in the hour lines and the hour numerals, so that the time can be read from the shadow on the sundial face, you must select a place to use the sundial outdoors. That is the topic of the next chapter.

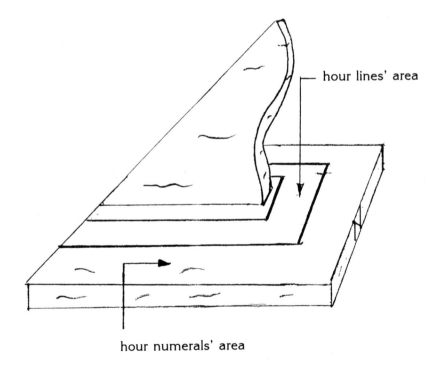

hour lines' area

hour numerals' area

Diagram 20

# Chapter 3

# Selecting a Site for the Sundial

Before putting in the lines for the hours on the sundial, you must select a spot where the sundial will always be positioned when in use at your home. Choose a sunny spot that is relatively unobstructed. Also, you must decide whether to mount the sundial permanently outdoors or simply select an exact spot where you will place it temporarily each time you use it. Either way will give good results.

Whichever way you select, it will be necessary for the sundial to be perfectly level. Check this with a bubble level. Also, the gnomon must point to true north. Find true north by using your adjustable magnetic compass or by aligning the gnomon in the direction of the North Star (see pp. 18-20).

If you will be using the sundial at a spot where it's never clamped down, like a sidewalk, do this: When it is correctly positioned, make two inconspicuous marks on the substratum around two corners of the sundial plate. Mark it with something that won't quickly be washed away by the weather. It will be adequate for many repositionings.

SIDEWALK - TOP VIEW

two marks for aligning bottom of plate for temporary use

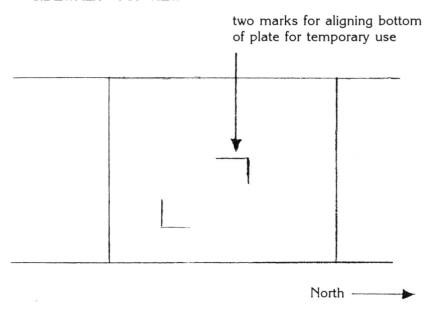

North ⟶

Diagram 21

If you are fastening this sundial to a railing, post, or other stationary structure, I recommend using woodscrews or C clamps. If you wish to mount the sundial permanently but have no existing rail or post on which to mount it, you can put in your own post.

Use a treated, outdoor post (4" x 4" x 6'). Put it in the ground, be sure it is level, then anchor with concrete. On top of it put a circular piece of wood with an 18 inch diameter (¾ inch plywood is adequate). Of course, it will require priming and painting to suit your wishes and to protect it

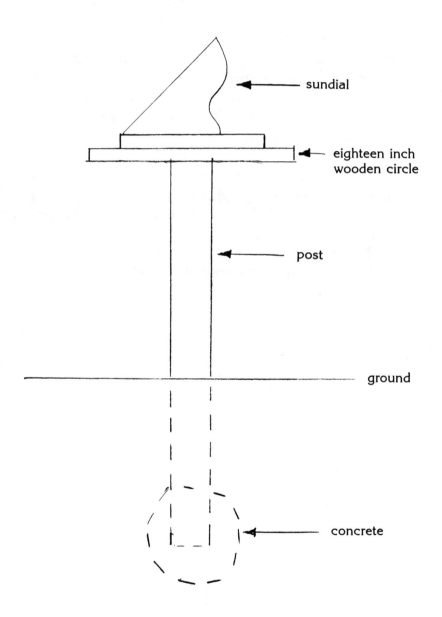

Diagram 22

from the elements. The circular piece of wood can be fastened to the post with two screws or by brackets attached to the sides of the post. Place the sundial on top of the circular piece. It is a good idea to mount the 18-inch diameter, wooden circle to the post before you anchor the post in concrete, because sometimes a woodscrew or lag screw can crack a wooden post.

Once you have secured the sundial in its place, whether permanently or temporarily, you are ready to finish it by marking and drawing in the lines representing the hours, half hours, and quarter hours. When you make the marks, the sundial must be in its chosen location, level, and pointing true north.

# Chapter 4

# Setting the Time on the Sundial

With the sundial in the correct position, you're ready to set the sundial time by marking shadow positions on the sundial face. Surprisingly though, to set the sundial time right, you must usually make it read a little different from clock time!

The reason for this is rooted in these scientific facts. A sundial works because the earth rotates. A sundial runs a little slow or fast compared to a clock because of the earth's tilt and its eccentric orbit of the sun. The exact number of minutes the sundial is fast or slow is known for each day of the year; it is always basically the same from year to year. These figures are listed in the chart included here called the *Equation of Time for Sundial Correction*. This chart tells you how fast or slow a sundial is because of the earth's tilt and its eccentric orbit of the sun. You will refer to the chart as you set the time on your sundial; also, you will use it to convert sundial time to clock time throughout the year.

**Equation of Time for Sundial Correction (+ means sundial fast; - means slow)**

|     | Jan. | Feb. | Mar. | Apr. | May | June | July | Aug. | Sept. | Oct. | Nov. | Dec. |
|-----|------|------|------|------|-----|------|------|------|-------|------|------|------|
| 1   | -3   | -14  | -13  | -4   | +3  | +2   | -4   | -6   | 0     | +10  | +16  | +11  |
| 2   | -4   | -14  | -12  | -4   | +3  | +2   | -4   | -6   | 0     | +10  | +16  | +11  |
| 3   | -4   | -14  | -12  | -4   | +3  | +2   | -4   | -6   | 0     | +11  | +16  | +10  |
| 4   | -5   | -14  | -12  | -3   | +3  | +2   | -4   | -6   | +1    | +11  | +16  | +10  |
| 5   | -5   | -14  | -12  | -3   | +3  | +2   | -4   | -6   | +1    | +11  | +16  | +10  |
| 6   | -6   | -14  | -12  | -3   | +3  | +2   | -4   | -6   | +1    | +12  | +16  | +9   |
| 7   | -6   | -14  | -11  | -2   | +3  | +1   | -5   | -6   | +2    | +12  | +16  | +9   |
| 8   | -6   | -14  | -11  | -2   | +4  | +1   | -5   | -6   | +2    | +12  | +16  | +8   |
| 9   | -7   | -14  | -11  | -2   | +4  | +1   | -5   | -6   | +2    | +13  | +16  | +8   |
| 10  | -7   | -14  | -11  | -2   | +4  | +1   | -5   | -5   | +3    | +13  | +16  | +7   |
| 11  | -8   | -14  | -10  | -1   | +4  | +1   | -5   | -5   | +3    | +13  | +16  | +7   |
| 12  | -8   | -14  | -10  | -1   | +4  | 0    | -5   | -5   | +3    | +13  | +16  | +7   |
| 13  | -8   | -14  | -10  | -1   | +4  | 0    | -6   | -5   | +4    | +14  | +16  | +6   |
| 14  | -9   | -14  | -10  | 0    | +4  | 0    | -6   | -5   | +4    | +14  | +16  | +6   |
| 15  | -9   | -14  | -9   | 0    | +4  | 0    | -6   | -5   | +5    | +14  | +15  | +5   |
| 16  | -10  | -14  | -9   | 0    | +4  | 0    | -6   | -4   | +5    | +14  | +15  | +5   |
| 17  | -10  | -14  | -9   | 0    | +4  | -1   | -6   | -4   | +5    | +14  | +15  | +4   |
| 18  | -10  | -14  | -8   | 0    | +4  | -1   | -6   | -4   | +6    | +15  | +15  | +4   |
| 19  | -11  | -14  | -8   | +1   | +4  | -1   | -6   | -4   | +6    | +15  | +15  | +3   |
| 20  | -11  | -14  | -8   | +1   | +4  | -1   | -6   | -4   | +6    | +15  | +15  | +3   |
| 21  | -11  | -14  | -7   | +1   | +4  | -1   | -6   | -3   | +7    | +15  | +14  | +2   |
| 22  | -11  | -14  | -7   | +1   | +4  | -2   | -6   | -3   | +7    | +15  | +14  | +2   |
| 23  | -12  | -14  | -7   | +2   | +3  | -2   | -6   | -3   | +7    | +16  | +14  | +1   |
| 24  | -12  | -13  | -7   | +2   | +3  | -2   | -6   | -3   | +8    | +16  | +13  | +1   |
| 25  | -12  | -13  | -6   | +2   | +3  | -2   | -6   | -2   | +8    | +16  | +13  | 0    |
| 26  | -12  | -13  | -6   | +2   | +3  | -3   | -6   | -2   | +8    | +16  | +13  | 0    |
| 27  | -13  | -13  | -6   | +2   | +3  | -3   | -6   | -2   | +9    | +16  | +13  | -1   |
| 28  | -13  | -13  | -5   | +2   | +3  | -3   | -6   | -1   | +9    | +16  | +12  | -1   |
| 29  | -13  |      | -5   | +3   | +3  | -3   | -6   | -1   | +9    | +16  | +12  | -2   |
| 30  | -13  |      | -5   | +3   | +3  | -3   | -6   | -1   | +10   | +16  | +12  | -2   |
| 31  | -13  |      | -4   |      | +3  |      | -6   | -1   |       | +16  |      | -3   |

When referring to the *Equation of Time for Sundial Correction* chart, find the month and day that are the same as the day you are using the chart. If the number listed there has a plus (+) in front of it, the sundial is that many minutes fast for that day. If the number has a minus (-) before it, the sundial is that many minutes slow for that day compared to a clock. Therefore, when you mark the position of the shadow to stand for an hour, half hour, or quarter hour position, you must do it this way.

If you are marking locations of the shadow on the sundial face on May 15, you would first note that the chart says a sundial for that day is four minutes fast. Therefore, when the clock says 9:56, you would mark the shadow's location by drawing a straight line along its edge across the time line area. That line would stand for 10:00.

TOP VIEW OF SUNDIAL

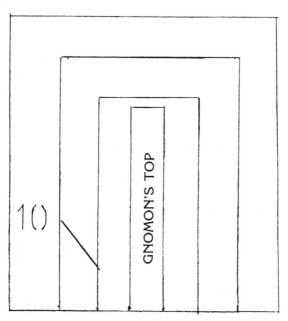

sundial is four minutes fast and
is marked accordingly

Diagram 23

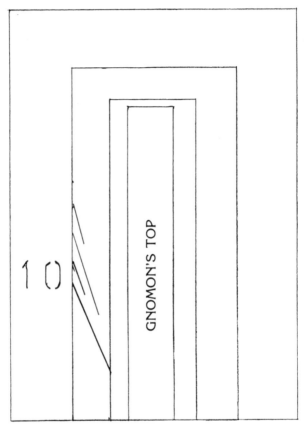

markings for the hour, half
hour, and quarter hours
Diagram 24

In order for your sundial to read time in quarter-hour increments, you will have to mark it
every fifteen minutes. So on the same day (May
15 in our example), when the clock says 10:11,
you would mark the shadow's location, but this
time draw a line just halfway across the time line
area. This line indicates that it is a quarter past
10:00, or 10:15. At 10:26 you would mark the
shadow's position for 10:30 by drawing a line $\frac{3}{4}$
of the way across the time line area. At 10:41

you would again draw a line going halfway across the time line area to indicate the next quarter-hour division of your timepiece. At 10:56 you would draw a full line along the shadow's edge to mark the dial for 11:00, and so on.

It is unlikely that you will put in all the time positions on the same day. Therefore, if you are putting more time positions in on another day, you will need to check the chart again to see if the sundial correction has changed for that day.

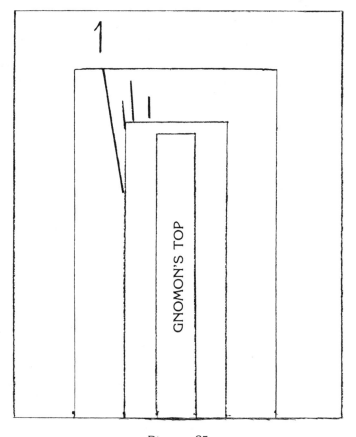

Diagram 25

If it was a little later in the year—for example, June 30—you would find from the chart that a sundial for that day is three minutes slow. Therefore, when the clock says 1:03, you would mark the shadow's location for 1:00. At 1:18 clock time you'd mark the shadow for 1:15. At 1:33 the shadow's location would stand for 1:30, and at 1:48, it would stand for 1:45.

When you are actually ready to mark shadow locations, it is best to use the following procedure. Instead of immediately drawing lines to stand for the time, first just place dots with a pen or pencil along the shadow's edge—something like what you did when you made the cardboard sundial. It's easy and takes just a

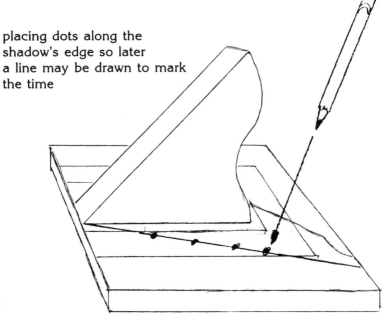

placing dots along the shadow's edge so later a line may be drawn to mark the time

Diagram 26

moment. Then when the sun is not glaring you can use a ruler or a straightedge to draw a line across the dots with a marker. Place four dots along the edge of the shadow for the hour, three for the half hour, and two for the quarter hour. When you draw the hour line, draw it all the way across the time line area. The half-hour lines go ¾ of the way across, and the quarter-hour lines go halfway across.

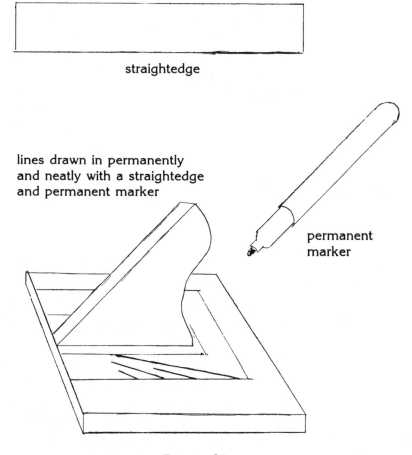

straightedge

lines drawn in permanently
and neatly with a straightedge
and permanent marker

permanent
marker

Diagram 27

## Standard Time, Daylight Saving Time, or Both?

When done marking the lines, you must decide how to number the marks. Do you want to number the marks to correspond to Standard Time or to Daylight Saving Time? Or, do you wish to be able to show both Standard Time and Daylight Saving Time on the sundial?

The simpler way to deal with this problem is to make the numbers correspond to Daylight Saving Time. The advantage to this is that Daylight Saving Time is in effect longer than Standard Time each year. Also, Daylight Saving Time occurs during the spring and summer when one is more likely to use a sundial. During that time the days are warm and there are more hours of sunlight each day than you have during Standard Time. Of course, you can still use the sundial in the winter. To convert a shadow reading to clock time, you simply subtract one hour. Then you make the *Equation of Time* chart correction.

You can also make the sundial with two separate sets of hour numbers: one for Daylight Saving Time and another for Standard Time. Do it this way. Put numbers next to the hour marks on the sundial face that correspond to Daylight Saving Time. Then make a U-shaped piece of wood from ¼ inch plywood or some other ¼ inch wood that fits exactly over the area for the hour numbers. Put hour numbers on it so that when it is mounted on the sundial face, the numbers will agree with Standard Time. If you want the Standard Time piece to be secure, drill four holes through the U-shaped piece of wood and corresponding holes in the sundial face to accommodate four woodscrews.

No. 8 ½ inch woodscrews

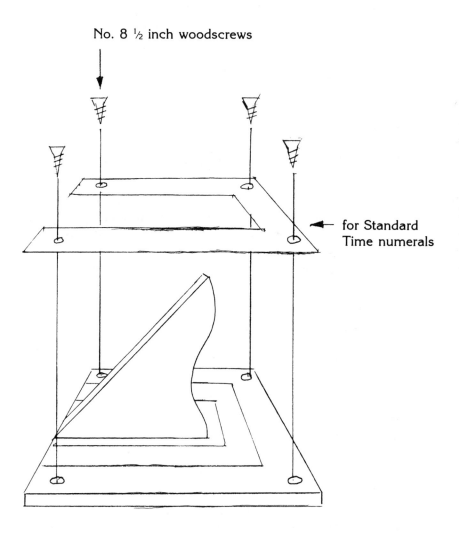

for Standard
Time numerals

Diagram 28

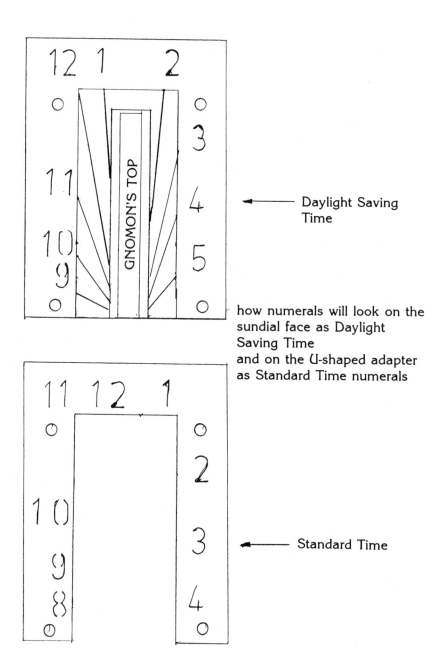

how numerals will look on the
sundial face as Daylight
Saving Time
and on the U-shaped adapter
as Standard Time numerals

Diagram 29

If you use a Standard Time adapter with the sundial during Daylight Savings Time, leave the screws in the holes in the sundial face. This will keep out rain, dew, and other forms of moisture.

When you have reached this point in making the sundial, there is one more thing you may want to do. Give the sundial another coat or two of clear polyurethane, especially if you intend it to be permanently mounted outdoors.

Whenever you use the sundial to tell time, remember that to get exact time you must do two things. First, note the time indicated by the shadow's location. Second, if there's a correction to make for that day, add or subtract it from your sundial time. In practice, if you just want approximate time, there is no need to refer to the sundial correction chart, especially during the summer months when the sundial averages only a few minutes different from clock time.

Besides having built an accurate timepiece, you've also built a scientific instrument. In addition to telling time, the position of the shadow on the sundial face on any day is evidence of the earth's motion. It is visible proof that the earth rotates, tilts, and orbits the sun in an eccentric path—varying its distance from the sun and its orbital speed—and obeying the laws that all satellites obey. Once you realize what you're really seeing when you look at a sundial's shadow, it's quite a sight!

# Chapter 5

# The Sundial As a Moondial

There are other things you can do with this sundial. You can observe the moon's shadow on the sundial face during the night just as you would observe the sun's shadow on the sundial during the day. From your observations, you can see clear evidence that the moon orbits the earth in a counterclockwise direction. Also, you can take a sundial along on trips in your car and observe changes that occur in the position of the shadow on the sundial face. From this, you can learn some interesting things about time zones.

If your sundial is permanently mounted, you might be thinking that it would be nice to have an extra sundial to take in your car to make observations about time zones. I suggested earlier that you save the wood you had left over from making the first sundial because you might want to make some more. There's obviously an advantage to not having to move a sundial that's permanently mounted. Also, you might prefer a separate sundial for observing the moon. One painted white with black numbers and lines is easier to read at night. Be assured, though, that the one sundial you've made is quite adequate to perform the following activities.

One of these activities is to observe the shadow of the moon on the sundial face during the cycle of the full moon. It's an easy and interesting way to watch the moon orbit the earth. This is how you do it. Wait for a moonlit night when the moon is casting a readable shadow on the sundial face (the moon need not be completely full). When the shadow reaches any of the quarter hour intervals on the sundial face (the shadow should actually touch the line), record two times. Write down what the sundial time is and what the clock says at the same instant. For example:

| Sundial Time | Clock Time |
|:---:|:---:|
| 8:45 | 7:39 |

Exactly one hour later, according to the clock, record both times again. You'll get results something like this:

| Sundial Time | Clock Time |
|:---:|:---:|
| 9:43 | 8:39 |

Repeat this procedure hour after hour until you go to bed. You'll get results like these:

| Sundial Time | Clock Time |
|:---:|:---:|
| 10:41 | 9:39 |
| 11:39 | 10:39 |
| 12:37 | 11:39 |

It's easy to see that each hour the sundial is losing about two minutes by moon time. This happens because the moon orbits the earth in a counterclockwise direction. The rotation of the earth causes the moon's shadow to mark the passing of time at about the same rate that time passes on a clock. But the moon's counterclockwise orbit of the earth causes the sundial reading to show a small loss of time hour after hour. This goes on day and night, all the time. Indeed, you'll see this the next night when the moon has risen high enough to cast a shadow on the sundial face. The moon's shadow will show the loss of about an hour compared to your first reading from the previous evening. Two days later the moon's shadow will have lost about two hours compared to the first night of your observations. This is why the cycle of the lunar month is always more than one day shorter than a solar month. The lunar month, which is the period between two successive new moons, averages 29 days, 12 hours, 44 minutes, and 2.8 seconds.

Another interesting thing to do with the sundial is to take it along in your car when you go on a trip. Also take along a magnetic compass and a bubble level. Whenever you've traveled about 100 miles or more, stop and take a reading of sundial time and compare it to clock time. Point the sundial north according to the direction shown by the compass. Also be sure that the sundial is level.

Here is what you'll see. As you travel north or south, you won't notice much difference between sundial time and clock time. There will be small differences, of course, because your gnomon may not be pointing exactly to true north or because the latitude changes. But traveling east, you'll see this. The sundial will slowly but surely gain time. The farther you go, the more time it will gain. Traveling west, the sundial will slowly but surely lose time. Why? These phenomena and other things will be explained in the next chapter.

# Chapter 6

# Basic Concepts

At the end of the first chapter, I suggested you save the small cardboard sundial you made. Now you can use it to get a better understanding of some concepts about the earth, the moon, and time zones.

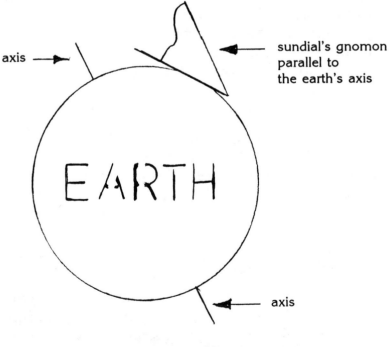

Diagram 30

When you made the cardboard sundial, you may have wondered why it had to point to true north. Also, you may have wondered why the gnomon had to have an angle with the same number of degrees as your latitude. The reason the gnomon must point to true north and have the same number of degrees in its angle as your latitude is to make the top of the gnomon exactly parallel to the earth's axis, which runs through the north and south poles.

If there actually were a real pole sticking straight up at the North Pole, it would be a good sundial. All you would need to do is add twenty-four hour line-marks around the pole. The shadow cast by the pole would then keep good time. The sundial you've made is something like a little North Pole positioned at a different spot on earth—your spot. Because its gnomon is perfectly parallel to the earth's axis, it keeps time much the same as the North Pole would if it were a sundial.

## A Science Fair Project

This and some other ideas will become clearer if we give a few more examples. In fact, you can use these examples, if you wish, as the basis for a science fair project to demonstrate how the earth's rotation and the sun affect the time of day and create the need for time zones. For this project you will need a world globe, a flashlight, and some masking tape.

First, put in some hour numerals on the cardboard sundial's face. Space them evenly and put them in clockwise order. Just put in even numerals as in the diagram.

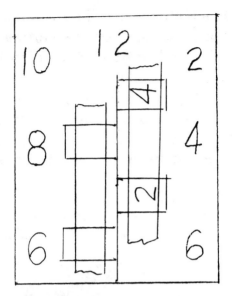

numerals added to
the sundial face
for understanding the
activities in
this chapter

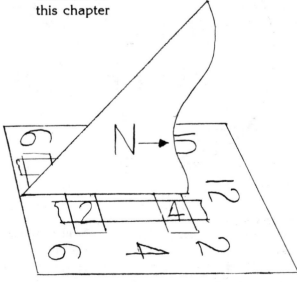

Diagram 31

Second, attach the cardboard sundial to the globe at about your spot on the earth with masking tape with the gnomon pointing to the North Pole. The tape will hold the sundial in place on the globe while you make some observations. One way of attaching the sundial with tape is to make a little ring out of a two-inch length of masking tape. Make it so the sticky part is on

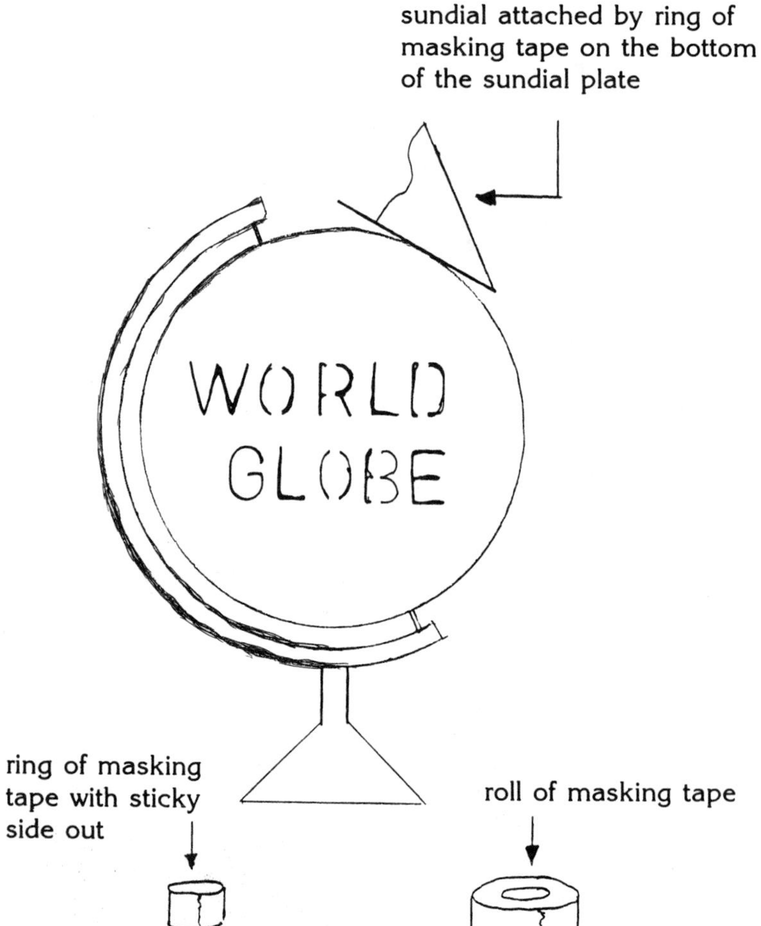

sundial attached by ring of masking tape on the bottom of the sundial plate

WORLD GLOBE

ring of masking tape with sticky side out

roll of masking tape

Diagram 32

the outside. Then stick the tape to the bottom of the cardboard sundial and place the sundial on the globe.

With the cardboard sundial attached to the world globe, we can get a better idea of what we mean when we say the sundial works because the earth rotates. Here is what to do. Shine the flashlight directly at the globe as you turn the globe slowly counterclockwise (west to east). You'll notice the shadow on the cardboard sundial moving clockwise, exactly the way it moves across your permanent sundial face. Indeed, the moving shadow on the globe brings this earth model to life. It illustrates the dynamics of a sundial is a quietly dramatic way.

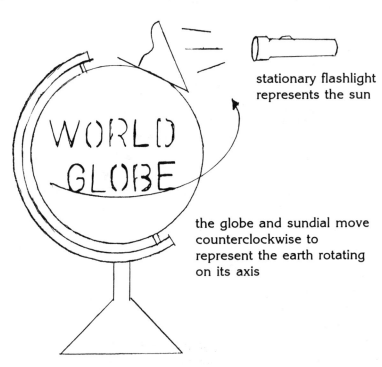

stationary flashlight represents the sun

the globe and sundial move counterclockwise to represent the earth rotating on its axis

Diagram 33

You can use these same objects to understand better how the moon's orbit of the earth causes the sundial to lose time. Leave the globe and the sundial stationary. Let the flashlight represent the moon. Move the flashlight counterclockwise to simulate the moon's counterclockwise orbit. Notice what happens to the shadow on the stationary sundial. It moves backwards, or counterclockwise, on the sundial face, as if time were running backwards. That is why the moon's shadow

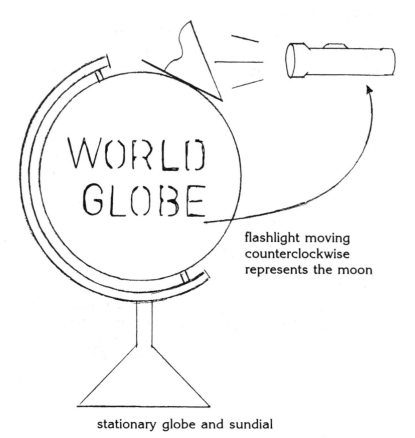

flashlight moving counterclockwise represents the moon

stationary globe and sundial

Diagram 34

slowly but surely loses time on the sundial. When you watch the moon's shadow on the sundial face, you are seeing the combined effects of two different motions of two different celestial bodies. The earth's rotation makes the moon shadow move rapidly clockwise, while at the same time the moon's orbit makes the shadow move slowly counterclockwise.

Use these same objects once more, this time to see how time zones work. You know that if you travel east with a sundial, you will see it gain time; if you travel west with a sundial, you will see it lose time. If you are traveling north or south with a sundial, you find that it gives readings that are close to clock time. You can see all this happen more easily and quickly with the globe, flashlight, and cardboard sundial.

Keep the globe stationary. Place the sundial properly (i. e., with the gnomon parallel to the earth's axis) on the East Coast of the United States. Shine the light on it. Notice the location of the shadow's position on the sundial face. Without rotating the globe, gently lift the cardboard sundial off its spot on the East Coast and move it to a spot on the West Coast where it is parallel to the North Pole. Notice the shadow now. You will see that it has lost time. Now move the cardboard sundial back to the East Coast without rotating the globe. Notice the shadow. Of course, it has regained the time. Now move the sundial north and south along the same meridian in the Northern Hemisphere without rotating the globe. Notice that the shadow stays about the same on the sundial face.

Basically, all this happens because the earth rotates from west to east. The time is constantly later to the east and earlier to the west. Yet, within a time zone all clocks are set to read the same as a matter of convenience and necessity in the modern world.

Besides being able to read correct time, you can learn a lot about the interrelation of the sun, the earth, the moon, and time by using a sundial. Earlier I predicted you would find the modern sundial educational, fascinating, and fun. I hope that this has been your experience.

# Notes

# Notes

# Notes

# Notes